YOUR KNOWLEDGE HAS VALUE

- We will publish your bachelor's and master's thesis, essays and papers

- Your own eBook and book - sold worldwide in all relevant shops

- Earn money with each sale

Upload your text at www.GRIN.com
and publish for free

John Bredakis

The proof by contradiction of the negation of Riemann Hypothesis

GRIN Verlag

Bibliografische Information der Deutschen Nationalbibliothek:

Die Deutsche Bibliothek verzeichnet diese Publikation in der Deutschen National-
bibliografie; detaillierte bibliografische Daten sind im Internet über http://dnb.d-
nb.de/ abrufbar.

Dieses Werk sowie alle darin enthaltenen einzelnen Beiträge und Abbildungen
sind urheberrechtlich geschützt. Jede Verwertung, die nicht ausdrücklich vom
Urheberrechtsschutz zugelassen ist, bedarf der vorherigen Zustimmung des Verla-
ges. Das gilt insbesondere für Vervielfältigungen, Bearbeitungen, Übersetzungen,
Mikroverfilmungen, Auswertungen durch Datenbanken und für die Einspeicherung
und Verarbeitung in elektronische Systeme. Alle Rechte, auch die des auszugsweisen
Nachdrucks, der fotomechanischen Wiedergabe (einschließlich Mikrokopie) sowie
der Auswertung durch Datenbanken oder ähnliche Einrichtungen, vorbehalten.

Imprint:

Copyright © 2013 GRIN Verlag GmbH
Druck und Bindung: Books on Demand GmbH, Norderstedt Germany
ISBN: 978-3-656-42219-8

This book at GRIN:

http://www.grin.com/en/e-book/213699/the-proof-by-contradiction-of-the-negation-
of-riemann-hypothesis

GRIN - Your knowledge has value

Der GRIN Verlag publiziert seit 1998 wissenschaftliche Arbeiten von Studenten, Hochschullehrern und anderen Akademikern als eBook und gedrucktes Buch. Die Verlagswebsite www.grin.com ist die ideale Plattform zur Veröffentlichung von Hausarbeiten, Abschlussarbeiten, wissenschaftlichen Aufsätzen, Dissertationen und Fachbüchern.

Visit us on the internet:

http://www.grin.com/

http://www.facebook.com/grincom

http://www.twitter.com/grin_com

The proof by contradiction of the negation of Riemann Hypothesis

Riemann Hypothesis $s=(\sigma+i.t)$

All the non trivial zeros of the Zeta function $\zeta(s)$ in the critical strip (0<Real s<1) are located on the critical line (Real s=1/2)

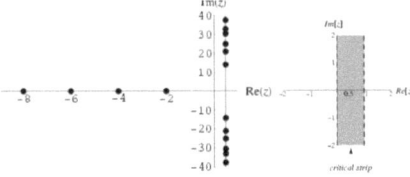

By my effort to solve:

   The greatest unsolved problem in Mathematics

I found that:

The Riemann Hypothesis is not valid

Iff there exist p, q, t, ξ ($p \# q$) to match this equation

$$\begin{array}{c} \text{Regardless of whether } p>q \text{ or } p<q \\ 0<p<1 \quad 0<q<1 \quad q=1-p \quad \boxed{-1 \leq \kappa < 0 \text{ for } t<0} \\ R = \tan(\kappa \cdot \frac{\pi}{2}) = \tan\left[\log(\xi) \cdot \frac{t}{2}\right] = \frac{\left[\xi^{p/2} + \xi^{q/2}\right]}{\left[\xi^{p/2} - \xi^{q/2}\right]} \cdot (p-q) \cdot \frac{t}{t^2 + p.q} \\ 0<\kappa \leq 1 \quad \xi>1 \end{array}$$

The above equation is correct:

$$\text{For } t>0 \text{ and } \boxed{M.t = t^2 + p.q} \quad M = \left[\frac{\kappa.\pi}{\log(\xi)} + (p.q) \cdot \frac{\log(\xi)}{\kappa.\pi}\right]^{0<\kappa\leq 1}$$

Those distinct t's (Roots of $t^2 - M.t + p.q = 0$) exist, but cannot be defined, since the exact value of ξ ($\xi>1$) is not known

http://Mathhighways.blogspot.com/

John Bredakis

Athens Greece 2013

A fact for the Zeta function ζ(s) s=(σ+i.t)

$$\theta(x) = \sum_{n=-\infty}^{n=+\infty} e^{-n^2 \cdot \pi \cdot x} \quad \Big| \quad \frac{\theta(x)-1}{2} = \sum_{n=1}^{+\infty} e^{-n^2 \cdot \pi \cdot x} \quad \Big| \quad \frac{\theta(x)}{\theta(1/x)} = \frac{1}{\sqrt{x}}$$

$$\pi^{-s/2} \cdot \Gamma(s/2) \cdot \zeta(s) = \pi^{-(1-s)/2} \cdot \Gamma[(1-s)/2] \cdot \zeta(1-s)$$

$$= \frac{-1}{s \cdot (1-s)} + \int_1^{+\infty} \left[x^{s/2} + x^{(1-s)/2} \right] \cdot x^{-1} \cdot \frac{\theta(x)-1}{2} \cdot dx$$

Checking the Riemann Hypothesis in the critical strip
s=(p±i.t) or s=(q±i.t) 0<p<1 0<q<1 q=1-p

I found a formula for A and B

$$\begin{aligned}
&= \pi^{-\left[\frac{p+i.t}{2}\right]} \cdot \Gamma\left[\frac{p+i.t}{2}\right] \cdot \zeta(p+i.t) = A \quad 0<p<1 \\
&= \pi^{-\left[\frac{q-i.t}{2}\right]} \cdot \Gamma\left[\frac{q-i.t}{2}\right] \cdot \zeta(q-i.t) = A \quad 0<q<1
\end{aligned} \quad \boxed{q=1-p}$$

$$\begin{aligned}
&= \pi^{-\left[\frac{p-i.t}{2}\right]} \cdot \Gamma\left[\frac{p-i.t}{2}\right] \cdot \zeta(p-i.t) = B \quad 0<p<1 \\
&= \pi^{-\left[\frac{q+i.t}{2}\right]} \cdot \Gamma\left[\frac{q+i.t}{2}\right] \cdot \zeta(q+i.t) = B \quad 0<q<1
\end{aligned} \quad \boxed{q=1-p}$$

And then I made a Working Hypothesis
ie: under what conditions A=B=0
expecting that this zero belongs to ζ(s)

For the proof of the formulas for A and B
see text

For the proof of the formulas for A and B
see text

$$\Bigg|\begin{matrix} = \pi^{-\left[\frac{p+i.t}{2}\right]} . \Gamma\left[\frac{p+i.t}{2}\right] . \zeta(p+i.t) = A & 0<p<1 \\ = \pi^{-\left[\frac{q-i.t}{2}\right]} . \Gamma\left[\frac{q-i.t}{2}\right] . \zeta(q-i.t) = A & 0<q<1 \end{matrix} \qquad \boxed{q=1-p}$$

$$A = \frac{-\left[t^2+p.q\right] - i.(p-q).t}{\left[t^2+p.q\right]^2 + (p-q)^2.t^2}$$

$$+ \int_1^{+\infty} \left[x^{p/2} + x^{q/2}\right].x^{-1}.\cos\left[\log(x).\frac{t}{2}\right].\frac{\theta(x)-1}{2}.dx$$

$$+ i. \int_1^{+\infty} \left[x^{p/2} - x^{q/2}\right].x^{-1}.\sin\left[\log(x).\frac{t}{2}\right].\frac{\theta(x)-1}{2}.dx$$

$$\Bigg|\begin{matrix} = \pi^{-\left[\frac{p-i.t}{2}\right]} . \Gamma\left[\frac{p-i.t}{2}\right] . \zeta(p-i.t) = B & 0<p<1 \\ = \pi^{-\left[\frac{q+i.t}{2}\right]} . \Gamma\left[\frac{q+i.t}{2}\right] . \zeta(q+i.t) = B & 0<q<1 \end{matrix} \qquad \boxed{q=1-p}$$

$$B = \frac{-\left[t^2+p.q\right] + i.(p-q).t}{\left[t^2+p.q\right]^2 + (p-q)^2.t^2}$$

$$+ \int_1^{+\infty} \left[x^{p/2} + x^{q/2}\right].x^{-1}.\cos\left[\log(x).\frac{t}{2}\right].\frac{\theta(x)-1}{2}.dx$$

$$- i. \int_1^{+\infty} \left[x^{p/2} - x^{q/2}\right].x^{-1}.\sin\left[\log(x).\frac{t}{2}\right].\frac{\theta(x)-1}{2}.dx$$

My Working Hypothesis
ie: under what conditions A=B=0
expecting that this zero belongs to ζ(s)

Suppose that:

$$M = \frac{\left[t^2 + p \cdot q\right]}{\left[t^2 + p \cdot q\right]^2 + (p-q)^2 \cdot t^2} = I\cos \qquad \begin{array}{l} 0 < p < 1 \\ q = 1-p \\ 0 < q < 1 \end{array} \qquad M = I\cos$$

$$I\cos = \int_1^{+\infty} \left[x^{p/2} + x^{q/2}\right] \cdot x^{-1} \cdot \cos\left[\log(x) \cdot \frac{t}{2}\right] \cdot \frac{\theta(x)-1}{2} \cdot dx$$

And suppose also that:

$$N = \frac{(p-q) \cdot t}{\left[t^2 + p \cdot q\right]^2 + (p-q)^2 \cdot t^2} = I\sin \qquad N = I\sin$$

$$I\sin = \int_1^{+\infty} \left[x^{p/2} - x^{q/2}\right] \cdot x^{-1} \cdot \sin\left[\log(x) \cdot \frac{t}{2}\right] \cdot \frac{\theta(x)-1}{2} \cdot dx$$

$$I\cos = \int_1^{+\infty} \left[x^{p/2} + x^{q/2}\right] \cdot x^{-1} \cdot \cos\left[\log(x) \cdot \frac{t}{2}\right] \cdot \frac{\theta(x)-1}{2} \cdot dx$$

$$\frac{\theta(x)-1}{2} = \sum_{n=1}^{+\infty} e^{-n^2 \cdot \pi \cdot x}$$

$$I\sin = \int_1^{+\infty} \left[x^{p/2} - x^{q/2}\right] \cdot x^{-1} \cdot \sin\left[\log(x) \cdot \frac{t}{2}\right] \cdot \frac{\theta(x)-1}{2} \cdot dx$$

By my Working Hypothesis
$$A = B = 0$$

$$I\sin = \frac{(p-q)\cdot t}{\Delta} \quad \text{and} \quad I\cos = \frac{[t^2 + p\cdot q]}{\Delta} \qquad \Delta = \left[t^2 + p\cdot q\right]^2 + (p-q)^2 \cdot t^2$$

$$\text{ie:} \quad \frac{\Delta \cdot I\sin}{(p-q)\cdot t} = \frac{\Delta \cdot I\cos}{[t^2 + p\cdot q]} = 1 \qquad \begin{array}{l} 0 < p < 1 \\ 0 < q < 1 \end{array} \quad q = 1 - p$$

By my Working Hypothesis
The following integral vanishes (becomes zero)

$$\log(x)\cdot\frac{t}{2} = A \qquad \Delta = \left[t^2 + p\cdot q\right]^2 + (p-q)^2 \cdot t^2$$

$$\int_{1}^{+\infty} \Delta \cdot \left[\frac{\left[x^{p/2} - x^{q/2}\right]}{(p-q)\cdot t} \cdot \sin(A) - \frac{\left[x^{p/2} + x^{q/2}\right]}{t^2 + p\cdot q} \cdot \cos(A) \right] \cdot x^{-1} \cdot \frac{\theta(x) - 1}{2} \cdot dx$$

By the Mean Value Theorem of Integral Calculus
there must be a ξ > 1 so that

$$\frac{\left[\xi^{p/2} - \xi^{q/2}\right]}{(p-q)\cdot t} \cdot \sin\left[\log(\xi)\cdot\frac{t}{2}\right] = \frac{\left[\xi^{p/2} + \xi^{q/2}\right]}{t^2 + p\cdot q} \cdot \cos\left[\log(\xi)\cdot\frac{t}{2}\right]$$

or

Regardless of whether p > q or p < q

$$0 < p < 1 \quad 0 < q < 1 \quad q = 1 - p$$

$$R = \tan\left(\kappa\cdot\frac{\pi}{2}\right) = \tan\left[\log(\xi)\cdot\frac{t}{2}\right] = \frac{\left[\xi^{p/2} + \xi^{q/2}\right]}{\left[\xi^{p/2} - \xi^{q/2}\right]} \cdot (p-q) \cdot \frac{t}{t^2 + p\cdot q}$$

$$0 < \kappa \leq 1 \qquad \xi > 1 \qquad -1 \leq \kappa < 0 \text{ for } t < 0$$

So far we have that iff the formulas of A and B are correct
Then the following is also correct

$$R = \tan\left(\kappa \cdot \frac{\pi}{2}\right) = \tan\left[\log(\xi) \cdot \frac{t}{2}\right] = \frac{\left[\xi^{p/2} + \xi^{q/2}\right]}{\left[\xi^{p/2} - \xi^{q/2}\right]} \cdot (p-q) \cdot \frac{t}{t^2 + p \cdot q}$$

Regardless of whether $p>q$ or $p<q$

$0<p<1$, $0<q<1$, $q=1-p$, $0<\kappa \leq 1$, $\xi>1$, $-1 \leq \kappa < 0$ for $t<0$

The next step is to select a t to satisfy the above

$$\tan\left[\log(\xi) \cdot \frac{M \pm \sqrt{M^2 - 4 \cdot (p \cdot q)}}{4}\right] = \frac{\left[\xi^{p/2} + \xi^{q/2}\right]}{\left[\xi^{p/2} - \xi^{q/2}\right]} \cdot (p-q) \cdot M^{-1}$$

$\xi>1$, $0<p<1$, $0<q<1$, $q=1-p$

Or

$$\tan(\phi) = \frac{\left[\xi^{[2p-1]/2} + 1\right]}{\left[\xi^{[2p-1]/2} - 1\right]} \cdot (p-q) \cdot M^{-1}$$

Working on the left part of the above equation

$\xi>1$ \hspace{3cm} $0<\kappa \leq 1$

$$\log(\xi) = 2\kappa \cdot \pi / \left[M + \sqrt{M^2 - 4 \cdot (p \cdot q)}\right] \quad \text{Or} \quad \log(\xi) = 2\kappa \cdot \pi / [M - T]$$

$-----=T-----$

$$\xi = e^{2\kappa \cdot \pi / [M + T]} \quad \text{or} \quad \xi = e^{2\kappa \cdot \pi / [M - T]}$$

$$0<p<1 \quad 0<q<1 \quad q=1-p \qquad \text{Regardless of whether } p>q \text{ or } p<q \qquad \boxed{-1\leq \kappa<0 \text{ for } t<0}$$

$$R = \tan(\kappa \cdot \frac{\pi}{2}) = \tan\left[\log(\xi)\cdot\frac{t}{2}\right] = \frac{\left[\xi^{p/2} + \xi^{q/2}\right]}{\left[\xi^{p/2} - \xi^{q/2}\right]} \cdot (p-q) \cdot \frac{t}{t^2 + p \cdot q}$$

$$0<\kappa \leq 1 \qquad \xi>1$$

Working on the right part of the above equation

$$0<p<1 \quad 0<q<1 \quad q=1-p \qquad\qquad\qquad\qquad\qquad\qquad \xi>1$$

$$\xi^{[2p-1]/2} = \frac{S+1}{S-1} \qquad \boxed{\frac{M \cdot R}{(p-q)} = \frac{M \cdot R}{(2p-1)} = S}$$

$$\xi = \left[\frac{S+1}{S-1}\right]^{2/[2p-1]}$$

$$\xi>1$$
$$\xi = \left[\frac{S+1}{S-1}\right]^{2/[2p-1]} = e^{\log\left[\frac{S+1}{S-1}\right]^{2/[2p-1]}}$$

$$\xi>1 \qquad\qquad 0<\kappa \leq 1$$
$$\xi = e^{2\kappa \cdot \pi / [M+T]} \quad \boxed{\text{or}} \quad \xi = e^{2\kappa \cdot \pi / [M-T]}$$

$$\overline{} = \log(\xi) \overline{}$$

$$\kappa = \frac{[M \pm T]}{2 \cdot \pi} \cdot \log\left[\frac{S+1}{S-1}\right]^{2/[2p-1]} \qquad \boxed{S = \frac{M \cdot R}{(2p-1)}} \qquad R = \tan(\kappa \cdot \frac{\pi}{2})$$

$$\boxed{0<\kappa \leq 1}$$

$$T = \sqrt{M^2 - 4 \cdot (p \cdot q)} \qquad 0<p<1 \quad 0<q<1 \quad q=1-p$$

$$0<\kappa \leq 1$$
$$\text{For } t>0 \text{ and } \boxed{M \cdot t = t^2 + p \cdot q} \quad M = \left[\frac{\kappa \cdot \pi}{\log(\xi)} + (p \cdot q) \cdot \frac{\log(\xi)}{\kappa \cdot \pi}\right]$$

Those distinct t's (Roots of $t^2 - M \cdot t + p \cdot q = 0$) exist, but cannot be defined, since the exact value of ξ ($\xi>1$) is not known

Those distinct roots are: $t_{1,2} = (M \pm T)/2$

$$\text{Regardless of whether } p>q \text{ or } p<q \quad \blacksquare \quad 0<p<1 \quad 0<q<1 \quad q=1-p \quad \xi>1$$

$$[p-q]=[2p-1] \hspace{5cm} [p-q]=[2p-1]$$

$$\xi^{[p-q]/2} = \left[\frac{S+1}{S-1}\right] = \left[\frac{M.R+(p-q)}{M.R-(p-q)}\right] \quad \blacksquare \quad \xi = \left[\frac{S+1}{S-1}\right]^{2/[2p-1]}$$

$$\xi^{[q-p]/2} = \left[\frac{S+1}{S-1}\right]^{-1} = \left[\frac{M.R+(p-q)}{M.R-(p-q)}\right]^{-1} = \left[\frac{M.R+(q-p)}{M.R-(q-p)}\right]$$

$$\xi^{[q-p]/2} = \left[\frac{M.R+(q-p)}{M.R-(q-p)}\right] \quad \blacksquare \quad \xi = \left[\frac{M.R+(q-p)}{M.R-(q-p)}\right]^{2/[2q-1]} \quad [2q-1]=[q-p]$$

Therefore in any case ξ>1 and close (to very close) to 1

As expected

Regardless of whether p>q or p<q $-1 \leq \kappa < 0$ for $t<0$

$0<p<1 \quad 0<q<1 \quad q=1-p$

$$R = \tan\left(\kappa \cdot \frac{\pi}{2}\right) = \tan\left[\log(\xi) \cdot \frac{t}{2}\right] = \frac{\left[\xi^{p/2} + \xi^{q/2}\right]}{\left[\xi^{p/2} - \xi^{q/2}\right]} \cdot (p-q) \cdot \frac{t}{t^2 + p.q}$$

$0<\kappa \leq 1 \quad \quad \xi>1$

For $t>0$ and $\boxed{M.t = t^2 + p.q}$ $\quad 0<\kappa \leq 1$

$$M = \left[\frac{\kappa.\pi}{\log(\xi)} + (p.q) \cdot \frac{\log(\xi)}{\kappa.\pi}\right]$$

Those distinct t's (Roots of $t^2 - M.t + p.q = 0$) exist, but cannot be defined, since the exact value of ξ (ξ>1) is not known

Those distinct roots are: t1,2 = (M±T)/2

The proof of the formulas for A and B

Checking the Riemann Hypothesis in the critical strip

$$\pi^{-s/2} \cdot \Gamma(s/2) \cdot \zeta(s) = \pi^{-(1-s)/2} \cdot \Gamma[(1-s)/2] \cdot \zeta(1-s)$$

$$= \frac{-1}{s \cdot (1-s)} + \int_{1}^{+\infty} \left[x^{s/2} + x^{(1-s)/2} \right] \cdot x^{-1} \cdot \frac{\theta(x)-1}{2} \cdot dx$$

$$\theta(x) = \sum_{n=-\infty}^{n=+\infty} e^{-n^2 \cdot \pi \cdot x} \quad \left| \quad \frac{\theta(x)-1}{2} = \sum_{n=1}^{+\infty} e^{-n^2 \cdot \pi \cdot x} \quad \right| \quad \frac{\theta(x)}{\theta(1/x)} = \frac{1}{\sqrt{x}}$$

A.

$$= \pi^{-\left[\frac{p+i \cdot t}{2}\right]} \cdot \Gamma\left[\frac{p+i \cdot t}{2}\right] \cdot \zeta(p+i \cdot t) = A \qquad 0<p<1$$

$$\boxed{q=1-p}$$

$$= \pi^{-\left[\frac{q-i \cdot t}{2}\right]} \cdot \Gamma\left[\frac{q-i \cdot t}{2}\right] \cdot \zeta(q-i \cdot t) = A \qquad 0<q<1$$

$$= \left[\frac{-1}{[p+i \cdot t] \cdot [q-i \cdot t]} = \frac{-1}{\left[t^2 + p \cdot q - i \cdot (p-q) \cdot t\right]} = \frac{-\left[t^2 + p \cdot q\right] - i \cdot (p-q) \cdot t}{\left[t^2 + p \cdot q\right]^2 + (p-q)^2 \cdot t^2} \right.$$

$$+ \int_{1}^{+\infty} \left[x^{(p+i \cdot t)/2} + x^{(q-i \cdot t)/2} \right] \cdot x^{-1} \cdot \frac{\theta(x)-1}{2} \cdot dx \Bigg]$$

Checking the Riemann Hypothesis in the critical strip

$$\pi^{-s/2} \cdot \Gamma(s/2) \cdot \zeta(s) = \pi^{-(1-s)/2} \cdot \Gamma[(1-s)/2] \cdot \zeta(1-s)$$

$$= \frac{-1}{s \cdot (1-s)} + \int_1^{+\infty} \left[x^{s/2} + x^{(1-s)/2} \right] \cdot x^{-1} \cdot \frac{\theta(x)-1}{2} \cdot dx$$

$$\theta(x) = \sum_{n=-\infty}^{n=+\infty} e^{-n^2 \cdot \pi \cdot x} \quad \left| \quad \frac{\theta(x)-1}{2} = \sum_{n=1}^{+\infty} e^{-n^2 \cdot \pi \cdot x} \quad \right| \quad \frac{\theta(x)}{\theta(1/x)} = \frac{1}{\sqrt{x}}$$

B.

$$= \pi^{-\left[\frac{p-i \cdot t}{2}\right]} \cdot \Gamma\left[\frac{p-i \cdot t}{2}\right] \cdot \zeta(p - i \cdot t) = B \quad 0<p<1$$

$$\boxed{q=1-p}$$

$$= \pi^{-\left[\frac{q+i \cdot t}{2}\right]} \cdot \Gamma\left[\frac{q+i \cdot t}{2}\right] \cdot \zeta(q + i \cdot t) = B \quad 0<q<1$$

$$= \left[\frac{-1}{[p-i \cdot t] \cdot [q+i \cdot t]} = \frac{-1}{\left[t^2 + p \cdot q + i \cdot (p-q) \cdot t\right]} = \frac{-\left[t^2 + p \cdot q\right] + i \cdot (p-q) \cdot t}{\left[t^2 + p \cdot q\right]^2 + (p-q)^2 \cdot t^2} \right.$$

$$\left. + \int_1^{+\infty} \left[x^{(p-i \cdot t)/2} + x^{(q+i \cdot t)/2} \right] \cdot x^{-1} \cdot \frac{\theta(x)-1}{2} \cdot dx \right]$$

Checking the Riemann Hypothesis in the critical strip

A+B

$$\begin{vmatrix} =\pi^{-\left[\frac{p+i.t}{2}\right]}.\Gamma\left[\frac{p+i.t}{2}\right].\zeta(p+i.t) = A & 0<p<1 \\ =\pi^{-\left[\frac{q-i.t}{2}\right]}.\Gamma\left[\frac{q-i.t}{2}\right].\zeta(q-i.t) = A & 0<q<1 \end{vmatrix} \quad \boxed{q=1-p}$$

$$\begin{vmatrix} =\pi^{-\left[\frac{p-i.t}{2}\right]}.\Gamma\left[\frac{p-i.t}{2}\right].\zeta(p-i.t) = B & 0<p<1 \\ =\pi^{-\left[\frac{q+i.t}{2}\right]}.\Gamma\left[\frac{q+i.t}{2}\right].\zeta(q+i.t) = B & 0<q<1 \end{vmatrix} \quad \boxed{q=1-p}$$

$$\boxed{\frac{A+B}{T}} = \frac{-2.\left[t^2+p.q\right]}{\left[t^2+p.q\right]^2+(p-q)^2.t^2} + \int_1^{+\infty}\left[x^{(p+i.t)/2}+x^{(p-i.t)/2}\right].x^{-1}.\frac{\Theta(x)-1}{2}.dx$$

$$+ \int_1^{+\infty}\left[x^{(q+i.t)/2}+x^{(q-i.t)/2}\right].x^{-1}.\frac{\Theta(x)-1}{2}.dx$$

$$\boxed{\frac{A+B}{T}} = \frac{-2.\left[t^2+p.q\right]}{\left[t^2+p.q\right]^2+(p-q)^2.t^2} + \int_1^{+\infty}\left[x^{p/2}+x^{q/2}\right].x^{-1}.\cos\left[\log(x).\frac{t}{2}\right].\frac{\Theta(x)-1}{1}.dx$$

Checking the Riemann Hypothesis in the critical strip

A-B

$$= \pi^{-\left[\frac{p+i.t}{2}\right]} \cdot \Gamma\left[\frac{p+i.t}{2}\right] \cdot \zeta(p+i.t) = A \quad 0<p<1$$

$$= \pi^{-\left[\frac{q-i.t}{2}\right]} \cdot \Gamma\left[\frac{q-i.t}{2}\right] \cdot \zeta(q-i.t) = A \quad 0<q<1$$

$\boxed{q=1-p}$

$$= \pi^{-\left[\frac{p-i.t}{2}\right]} \cdot \Gamma\left[\frac{p-i.t}{2}\right] \cdot \zeta(p-i.t) = B \quad 0<p<1$$

$$= \pi^{-\left[\frac{q+i.t}{2}\right]} \cdot \Gamma\left[\frac{q+i.t}{2}\right] \cdot \zeta(q+i.t) = B \quad 0<q<1$$

$\boxed{q=1-p}$

$$\boxed{\frac{A-B}{T}} = \frac{-2.i.(p-q).t}{\left[t^2+p.q\right]^2+(p-q)^2.t^2} + \int_1^{+\infty}\left[x^{(p+i.t)/2} - x^{(p-i.t)/2}\right] \cdot x^{-1} \cdot \frac{\theta(x)-1}{2} \, dx$$

$$- \int_1^{+\infty}\left[x^{(q+i.t)/2} - x^{(q-i.t)/2}\right] \cdot x^{-1} \cdot \frac{\theta(x)-1}{2} \, dx$$

$$\boxed{\frac{A-B}{T}} = \frac{-2.i.(p-q).t}{\left[t^2+p.q\right]^2+(p-q)^2.t^2} + i \cdot \int_1^{+\infty}\left[x^{p/2} - x^{q/2}\right] \cdot x^{-1} \cdot \sin\left[\log(x) \cdot \frac{t}{2}\right] \cdot \frac{\theta(x)-1}{1} \, dx$$

Checking the Riemann Hypothesis in the critical strip

By Combination

A =

$$= \pi^{-\left[\frac{p+i.t}{2}\right]} . \Gamma\left[\frac{p+i.t}{2}\right] . \zeta(p+i.t) = A \qquad 0<p<1$$

$$= \pi^{-\left[\frac{q-i.t}{2}\right]} . \Gamma\left[\frac{q-i.t}{2}\right] . \zeta(q-i.t) = A \qquad 0<q<1$$

$$\boxed{q=1-p}$$

$$= \pi^{-\left[\frac{p-i.t}{2}\right]} . \Gamma\left[\frac{p-i.t}{2}\right] . \zeta(p-i.t) = B \qquad 0<p<1$$

$$= \pi^{-\left[\frac{q+i.t}{2}\right]} . \Gamma\left[\frac{q+i.t}{2}\right] . \zeta(q+i.t) = B \qquad 0<q<1$$

$$\boxed{q=1-p}$$

$$A = \frac{-\left[t^2+p.q\right] - i.(p-q).t}{\left[t^2+p.q\right]^2 + (p-q)^2.t^2}$$

$$+ \int_1^{+\infty} \left[x^{p/2} + x^{q/2}\right] . x^{-1} . \cos\left[\log(x).\frac{t}{2}\right] . \frac{\Theta(x)-1}{2} . dx$$

$$+ i. \int_1^{+\infty} \left[x^{p/2} - x^{q/2}\right] . x^{-1} . \sin\left[\log(x).\frac{t}{2}\right] . \frac{\Theta(x)-1}{2} . dx$$

Checking the Riemann Hypothesis in the critical strip

By Combination

B=

$$= \pi^{-\left[\frac{p+i.t}{2}\right]} . \Gamma\left[\frac{p+i.t}{2}\right] . \zeta(p+i.t) = A \quad 0<p<1$$

$$= \pi^{-\left[\frac{q-i.t}{2}\right]} . \Gamma\left[\frac{q-i.t}{2}\right] . \zeta(q-i.t) = A \quad 0<q<1$$

$\boxed{q=1-p}$

$$= \pi^{-\left[\frac{p-i.t}{2}\right]} . \Gamma\left[\frac{p-i.t}{2}\right] . \zeta(p-i.t) = B \quad 0<p<1$$

$$= \pi^{-\left[\frac{q+i.t}{2}\right]} . \Gamma\left[\frac{q+i.t}{2}\right] . \zeta(q+i.t) = B \quad 0<q<1$$

$\boxed{q=1-p}$

$$B = \frac{-\left[t^2+p.q\right] + i.(p-q).t}{\left[t^2+p.q\right]^2 + (p-q)^2.t^2}$$

$$+ \int_1^{+\infty} \left[x^{p/2} + x^{q/2}\right] . x^{-1} . \cos\left[\log(x).\frac{t}{2}\right] . \frac{\Theta(x)-1}{2} . dx$$

$$- i. \int_1^{+\infty} \left[x^{p/2} - x^{q/2}\right] . x^{-1} . \sin\left[\log(x).\frac{t}{2}\right] . \frac{\Theta(x)-1}{2} . dx$$

Checking the Riemann Hypothesis in the critical strip
By Combination – Summary

$$= \pi^{-\left[\frac{p+i.t}{2}\right]} . \Gamma\left[\frac{p+i.t}{2}\right] . \zeta(p+i.t) = A \qquad 0<p<1$$

$$= \pi^{-\left[\frac{q-i.t}{2}\right]} . \Gamma\left[\frac{q-i.t}{2}\right] . \zeta(q-i.t) = A \qquad 0<q<1$$

$\boxed{q=1-p}$

$$= \pi^{-\left[\frac{p-i.t}{2}\right]} . \Gamma\left[\frac{p-i.t}{2}\right] . \zeta(p-i.t) = B \qquad 0<p<1$$

$$= \pi^{-\left[\frac{q+i.t}{2}\right]} . \Gamma\left[\frac{q+i.t}{2}\right] . \zeta(q+i.t) = B \qquad 0<q<1$$

$\boxed{q=1-p}$

$$A = \frac{-\left[t^2+p.q\right] - i.(p-q).t}{\left[t^2+p.q\right]^2 + (p-q)^2.t^2}$$

$$+ \int_1^{+\infty} \left[x^{p/2} + x^{q/2}\right].x^{-1}.\cos\left[\log(x).\frac{t}{2}\right].\frac{\Theta(x)-1}{2}.dx$$

$$+ i.\int_1^{+\infty} \left[x^{p/2} - x^{q/2}\right].x^{-1}.\sin\left[\log(x).\frac{t}{2}\right].\frac{\Theta(x)-1}{2}.dx$$

$$B = \frac{-\left[t^2+p.q\right] + i.(p-q).t}{\left[t^2+p.q\right]^2 + (p-q)^2.t^2}$$

$$+ \int_1^{+\infty} \left[x^{p/2} + x^{q/2}\right].x^{-1}.\cos\left[\log(x).\frac{t}{2}\right].\frac{\Theta(x)-1}{2}.dx$$

$$- i.\int_1^{+\infty} \left[x^{p/2} - x^{q/2}\right].x^{-1}.\sin\left[\log(x).\frac{t}{2}\right].\frac{\Theta(x)-1}{2}.dx$$

Some general remarks on the Zeta function ζ(s)
And a reference to my pdf
Understanding the Zeta function and the Riemann Hypothesis

Relation between the Zeta function and the primes

$$\zeta(s) = 1 + \frac{1}{2^s} + \frac{1}{3^s} + \frac{1}{4^s} + \frac{1}{5^s} + \cdots$$

$$\frac{1}{2^s}\zeta(s) = \frac{1}{2^s} + \frac{1}{4^s} + \frac{1}{6^s} + \frac{1}{8^s} + \frac{1}{10^s} + \cdots$$

Subtracting the second from the first we remove all elements that have a factor of 2:

$$\left(1 - \frac{1}{2^s}\right)\zeta(s) = 1 + \frac{1}{3^s} + \frac{1}{5^s} + \frac{1}{7^s} + \frac{1}{9^s} + \frac{1}{11^s} + \frac{1}{13^s} + \cdots$$

Repeating for the next term:

$$\frac{1}{3^s}\left(1 - \frac{1}{2^s}\right)\zeta(s) = \frac{1}{3^s} + \frac{1}{9^s} + \frac{1}{15^s} + \frac{1}{21^s} + \frac{1}{27^s} + \frac{1}{33^s} + \cdots$$

Subtracting again we get:

$$\left(1 - \frac{1}{3^s}\right)\left(1 - \frac{1}{2^s}\right)\zeta(s) = 1 + \frac{1}{5^s} + \frac{1}{7^s} + \frac{1}{11^s} + \frac{1}{13^s} + \frac{1}{17^s} + \cdots$$

where all elements having a factor of 3 or 2 (or both) are removed.

It can be seen that the right side is being sieved. Repeating infinitely we get:

$$\cdots \left(1 - \frac{1}{11^s}\right)\left(1 - \frac{1}{7^s}\right)\left(1 - \frac{1}{5^s}\right)\left(1 - \frac{1}{3^s}\right)\left(1 - \frac{1}{2^s}\right)\zeta(s) = 1$$

Dividing both sides by everything but the ζ(s) we obtain:

$$\zeta(s) = \frac{1}{\left(1 - \frac{1}{2^s}\right)\left(1 - \frac{1}{3^s}\right)\left(1 - \frac{1}{5^s}\right)\left(1 - \frac{1}{7^s}\right)\left(1 - \frac{1}{11^s}\right)\cdots}$$

General remarks on the Zeta function ζ(s)
And the first 42 roots in the critical strip

14.134725142	1	$\zeta(1) \to +\infty$
21.022039639	2	This sum can be expressed as:
25.010857580	3	a product of prime numbers $\zeta(3/2)=2.612$
30.424876126	4	(See Wikipedia)
32.935061588	5	$\zeta(2)=\pi^2/6 = 1.645$
37.586178159	6	
40.918719012	7	$\zeta(3) = 1.202$
43.327073281	8	$\zeta(s) = \sum_{n=1}^{+\infty} 1/n^s$
48.005150881	9	
49.773832478	10	$\zeta(4)=\pi^4/90 =1.0823$
52.970321478		
56.446247697		
59.347044003		The zeta function in the complex plane has:
60.831778525		
65.112544048		
67.079810529		At s=1 a singularity:
69.546401711		(A single pole with residue one)
72.067157674		
75.704690699		At s=0 the value of $-(1/2)$
77.144840069	20	
79.337375020		
82.910380854		At s=-1 the value of $-(1/12)$
84.735492981		
87.425274613		
88.809111208		Trivial zeros at s = -2, -4, -6, etc
92.491899271		
94.651344041		
95.870634228		The beauty of the zeta function is:
98.831194218		in the critical strip
101.317851006	30	0< Real s <1
103.725538040		with plenty of zeros
105.446623052		
107.168611184		The conjecture by Riemann is that:
111.029535543		all those zeros are in the line
111.874659177		Real s = (1/2)
114.320220915		
116.226680321		Notice that: $\zeta(1/2 + i.t) = \zeta(1/2 - i.t)$
118.790782866		
121.370125002		* Needless to say that the best
122.946829294	40	mathematical brains were
124.256818554	41	summond to develop
127.516683880	42	those supercomputer numbers

For further details see my pdf:
Understanding the Zeta function
and the Riemann Hypothesis

http://Mathhighways.blogspot.com/
John Bredakis
Athens Greece 2013

References

1. **Higher Mathematics for beginners:**
   by Ya.B.Zeldovich
   (Mir Publishers Moscow 1973)

2. **Calculus with analytic geometry:**
   by Harley Flanders and Justin J Price
   (Academic Press 1978)

3. **A brief course of higher mathematics:**
   by V.A.Kudryavtsev and B.P.Demidovich
   (Mir Publisher's Moscow 1980)

4. **Concice Encyclopedia of Mathematics:**
   by W.Gellert,H.Kustner,M Hellwich,H Kastner
   (Van Nostrand Reinhold Company New York and other cities 1977)

5. **Computational Mathematics:**
   by B.P Demitovich and I.A.Maron
   (Mir publishers Moscow 1976)

6. **Advanced calculus:**
   by Leopold Flatto
   (The Wiiliams and Wilkins Company - Baltimore 1982)

7. **Mathematics Handbook for Science and Engineering:**
   by: Royal Lennart Rade and Bertil Westegren
   Fifth edition - 2004
   Springer Verlag Publications Inc
   Berlin - Heidelberg - New York

8. **Mathematical methods for physicists and engineers:**
   by: Royal Eugene Collins - 2nd corrected edition
   Dover Publications Inc - Mineola New York - USA 1991

9. **Differential Equations:**
   A systems approach - by: Jack Goldberg - and Merle C.Potter
   Prentice Hall International Editions
   Upper Saddle River , NJ - USA - 1998

And a lot of personal work

I like to express my special thanks to Professor of Mathematics
Elias Kastanas
for his advise and check up

Additional References From The Internet

The Wikipedia , the free encyclopedia

An Introduction to the Riemann Hypothesis
An Introduction to the Riemann Hypothesis Theodore J. Yoder∗ May 27, 2011 Abstract Here we discuss the most famous unsolved problem in mathematics,
sections.maa.org/epadel/students/studentWinners/2011_Yoder.p ...

Zeros of the Riemann Zeta-function on the critical line
Universit a degli Studi ROMA TRE Zeros of the Riemann Zeta-function on the critical line Author:
Lorenzo Menici Supervisor: Prof. Francesco Pappalardi
www.mat.uniroma3.it/scuola_orientamento/alumni/laureati/meni ...

Riemann Zeta Function - ONID
Riemann Zeta Function Bent E. Petersen January 23, 1996 ... the students learned about the Riemann hypothesis { an important part of our mathematical heritage and culture.
people.oregonstate.edu/~peterseb/misc/docs/zeta.pdf

Numerical evaluation of the Riemann Zeta-function
PDF: Xavier Gourdon and Pascal Sebah
1. *Numerical evaluation of the Riemann. Zeta-function. Xavier Gourdon and Pascal* Sebah. July 23, 20031. We expose techniques that permit to approximate the ..

Odlyzko–Schönhage algorithm - Wikipedia, the free encyclopedia
In mathematics, the Odlyzko–Schönhage algorithm is a fast algorithm for evaluating the Riemann
zeta function at many points, introduced by (Odlyzko & Schönhage 1988).
en.wikipedia.org/wiki/Odlyzko-Schonhage

And my pdf: Big Bang in Math,John Bredakis Method & the Gamma function

The Gamma function is difficult to proove and easy to use under a proper guidance.

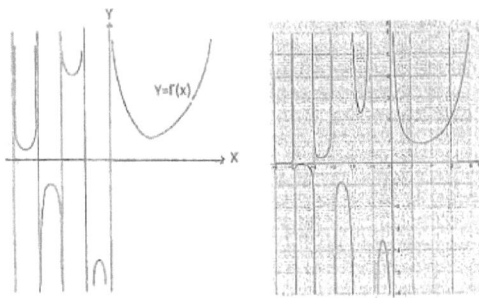

John.K.Bredakis MD

Assistant Professor University of Athens

American Board Certified Cardiologist

Born in Athens Greece 28/11/1946

Graduate of the medical school (1970)
University of Athens

Trained in internal medicine and Cardiology
(1970-1977)
Chicago - USA

Consultant Cardiologist - Areteion Hospital Athens Greece
Since 1977

Thanks God , uncle Fotis , Areteion Hospital
my parents , my wife Sofia
and professors C.Tountas , D.Voros , G.Limouris

A special thanks also to the professor Elias Kastanas
(Professor of mathematics - Engineer - Computer scientist etc)
The creator of my blog

I would also like to thank very much Professor of mathematics
Themistoklis Rassias

Athens Greece 2013

The reason to deal with the Zeta function
ζ(s), despite my shallow knowledge of complex analysis

The common things between those pdf of mine

Big Bang in Math , John Bredakis method & the Gamma function
and
The proof by contradiction of the negation of Riemann Hypothesis

The Gamma function and the tranfer of i from denominator to numerator

Trying to understand the difficult to comprehend topic of complex analysis I ended up to the calculus of residues.

And then I realized that the complex analysis can be bypassed to a large part by the classical advanced analysis , hence my pdf
An attempt to bypass the calculus of residues

Searching very carefully in the Internet , I found the one and only pdf
by **Theodore Yoder (Introduction to Riemann Hypothesis)**
unique in the sense of handling the ζ(s), with minimal complex analysis and epecially with the introduction of the idea of analytic continuation

Thereafter I needed only few lines from the pdf by Lorentzo Menici to comprehend the Riemann-Siegel formula , hence my pdf:
Understanding the Zeta function and the Riemann Hypothesis

This is the whole story – Handling Mathematics for many years

I must admitt that my mentor in Mathematics was my beloved Uncle Fotis , a medical doctor , brilliant mathematical brain and great teacher of Mathematics

Sincerelly:
John Bredakis MD
http://Mathhighways.blogspot.com/

Athens Greece 2013